Copyright © 2013 Stefan Burger. All rights reserved. No part of this publication may be reproduced, stored or transmitted, in any form or by any means without prior permission in writen.

© 2009 IEEE. Reprint, with permission, from: Wrapped Microstrip Antennas for Laptop Computers; Guterman, J.; Moreira, A. A.; Peixeiro, C.; Rahmat-Samii, Y.; IEEE Antennas and Propagation Magazine.

While the author believes that the information and guidance given in this work is correct, all parties must rely upon their own skill and judgement when making use of it. Neither the author nor the publisher assumes any liability to anyone for any loss or damage caused by error or omission in this work, whether such error or omission is the result of negligence or any other cause. Any and all such liability is disclaimed.

ISBN: 9783735723826

Herstellung und Verlag: BoD – Books on Demand, Norderstedt

Contents

Forward 3
Chapter 1 4
EMR/RF Basics 4
    Wavelength 4
    Magnetic Field 5
    Electric Field 7
    Alternating Power Source 8
    Power Delivered 13
    Maximum Exposure Level 17
Chapter 2 19
Radiation Systems In Our Lives 19
    Microwave Oven 19
    WLAN 2.4 GHz 21
    Cordless Phone 22
    Mobile Phone / Cell Phone 23
    Smart Meters 25
    Tablets and Notebooks 26
Chapter 3 29
How To Minimise Exposure 29
    Mobile Phones 29
    iPad & Tablet 30
    Microwave Ovens 31
    Baby Monitor 31
    Bluetooth Headset and Google Glass 31
    What doesn't work or is counter-productive: 32
    Cell phone radiation shield stickers 32
    Shielding cover 32
    Shielding pack 32
    Plug-in anti-radiation devices and stickers 33
Bibliography 34

Forward

Nowadays there are many devices that generate electro-magnetic radiation (EMR) and high, radio frequency (RF) radiation. The common systems are base stations, mobile phones, microwave ovens, tablets such as iPads, WiFi-enabled laptops, game consoles and smart meters.

This booklet explains some basics about electro-magnetic radiation so that you are able to guess whether a system could be dangerous for your health or not. Complex RF theory is explained clearly and simply. Chapter 1 outlines important background information on how RF radiation and electromagnetic radiation are created as well as when and why it can be dangerous for humans.

Chapter 2 shows measurement results from typical systems at home and in the office and an interpretation of them.

In Chapter 3 some suggestions for minimising electro-magnetic exposure are made.

Chapter 1

EMR/RF Basics

How is power transported without cables?

We need a minimum of two cables to get power to a device. Regardless of whether it is powered by a battery or the 240V power socket in the wall.
In the wires connected to your device, the electric current flows towards the device and back to the source. Some of the power delivered is used in your device. For example, a solder iron will become hot, or a Hi-Fi system will play music.

As the name "electro-magnetic waves" implies, the wave has two components: an electrical field and a magnetic field.

Wavelength

Imagine placing a pendulum on a car and driving down the road at a speed (**v**). If you then paint the position of the pendulum on the street while the car is moving, the resulting curve is a sinus curve and the length of one wave is the wavelength λ (Lamda).
We use the same calculation to calculate the wavelength of electro-magnetic waves. The speed here is the speed of light c_0=299,792,458 m/s. If an exact result is not required, we can use the rounded-up light speed: $c_0 \approx 300,00$ km/s.

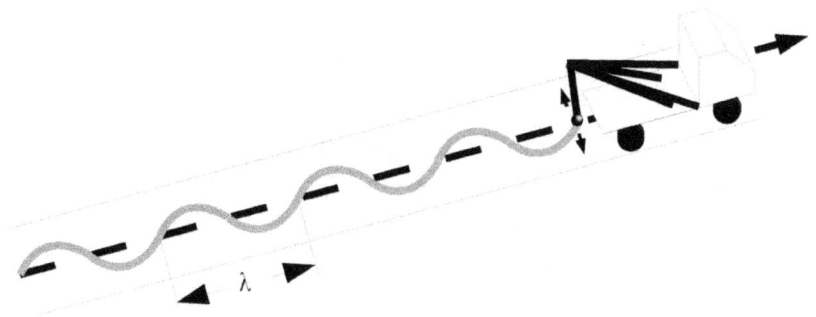

Image 1: Wavelength of an oscillation

Magnetic Field

Maybe you did this experiment at school:
If we connect a voltage source to a wire, a current will flow. Immediately around the wire is a magnetic field which travels at the speed of light to infinity. By increasing the distance, the measured magnetic strength is decreased because all the power is then spread over the surrounding space.

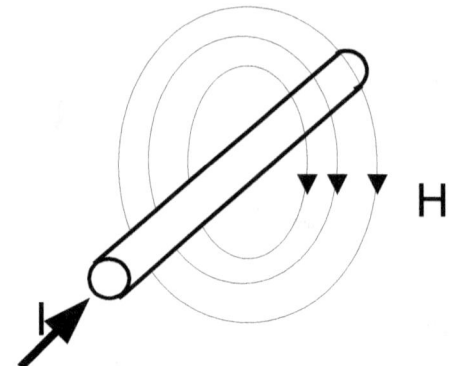

Image 2: Magnetic field around wire

This also works the opposite way: if we move a wire in a magnetic field, a current is driven through the wire. If the wire isn't moving, no current will be generated and no power will be transported. Instead of moving the wire, we could also change the intensity of the magnetic field. When the field changes, a current is generated.

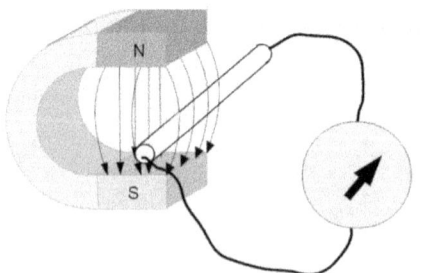

Image 3: Moving wire in magnetic field

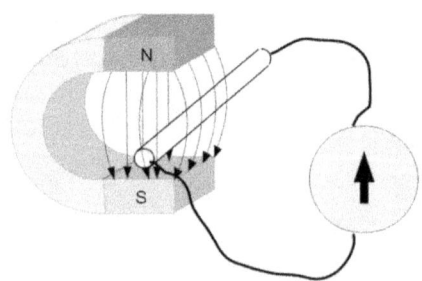

Image 4: Wire in magnetic field

Electric Field

If we connect two wires to a voltage source and place the open ends nearby, we can find an electric field between the open ends. Each field line is at a right angle to the conductor, starts at one side and ends on the other side of the open-ended wires. If we connect two plates to the open ends and move the plates so that they are finally one above the other, we can observe the field configuration you can see in image 5d. However, this field is connected to the plates and isn't moving and so no power is transported.

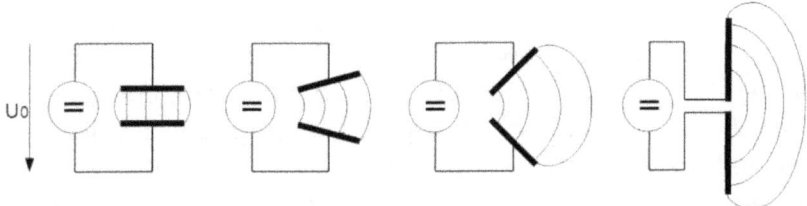

Image 5: Static Electric Field

To transport power through free space you need a changing field.

Alternating Power Source

An alternating voltage source is connected with two wires, they are one above the other. This configuration is the well-known dipole antenna.
If the voltage is increased, a current flows through the wires. The current flows at the speed of light, which is very fast, but it still takes time. Like cars at a traffic light: if the light changes to green the cars do not all start at once. One car has to move before the next car can start to move. The same happens with the electrons in the wire.

In Image 6 you can see how the current flows in the wire and also how the electric field travels outwards from the wire, both at the speed of light. Illustrated on top is the voltage wave in the connected line with progressing time.
In image 6d the electrons in the end of the wire are starting to travel and the voltage of the source has reached the maximum level. Now the current changes its direction and with it, the beginning of the electrical field travels back to the connection point. Simultaneously, the rest of the electrical field travels further out. Finally, at the bottom on the right in image 6, the field-lines join up to form a closed line. Unlike with the static field, where the electrical field is connected to the wire, we have an electrical field vortex which is no longer connected to the wire.
Image 7 shows the electrical and magnetic field around a dipole. The magnetic field is in concentric circles at a right-angle to the electrical field.

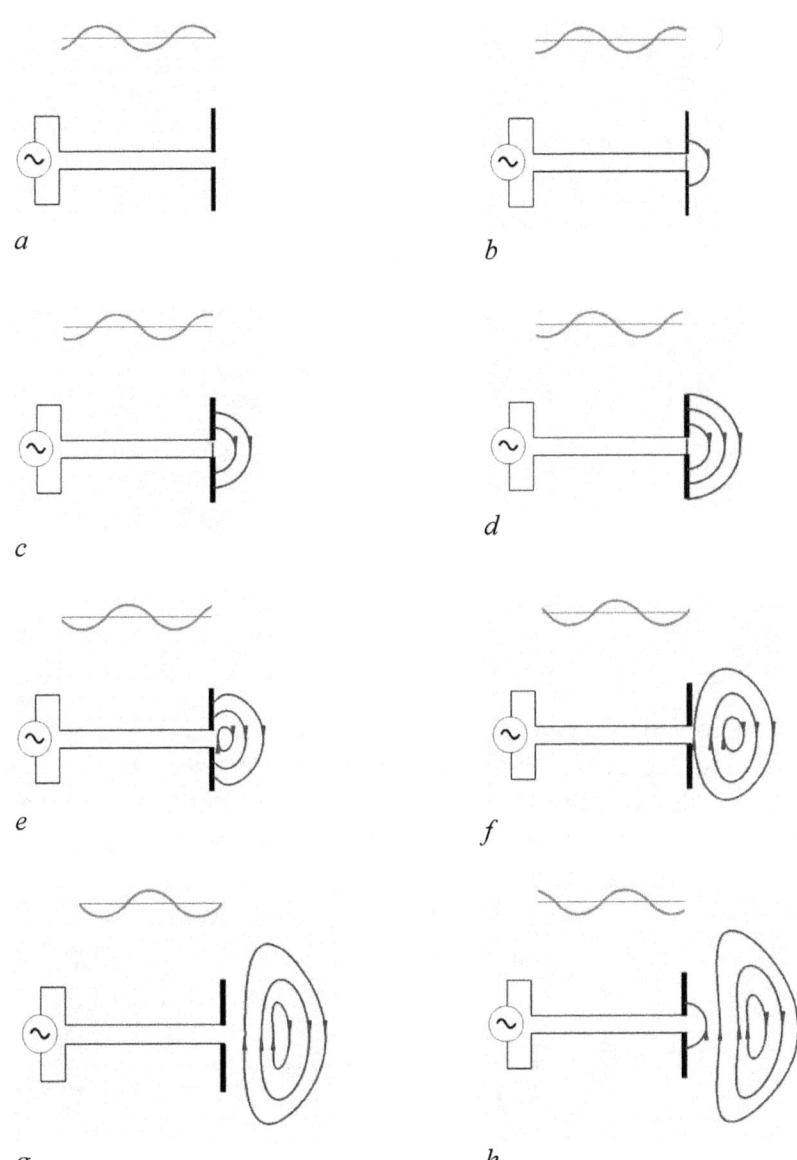

Image 6: Electric field at a dipole

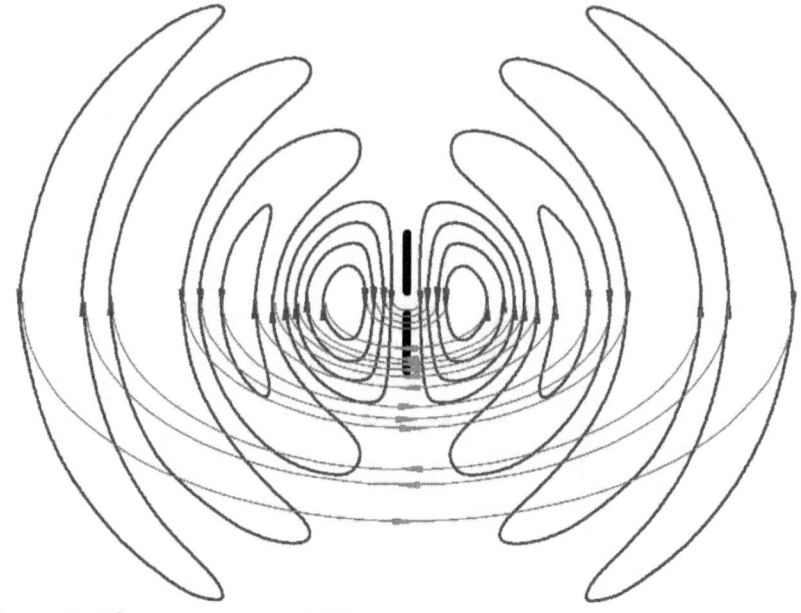

Image 7: Electro-magnetic field at an antenna

In the depiction above, the dipole length matches the wave length of the source. As maximum voltage is reached, the electrical field also reaches the end of the antenna. If an antenna is too short or too long, the electrical field can't form the closed lines perfectly, and the antenna doesn't work well.

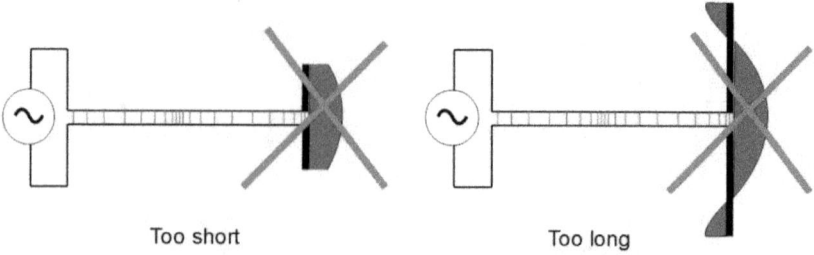

The energy delivered spreads out in space. This means the energy density decreases with distance from the antenna because the volume increases with distance.

In diagram 1 we see the power density over distance on a logarithmic scale. The power density drops down fast and reaches around 1% at 1 Lamda (λ) distance.

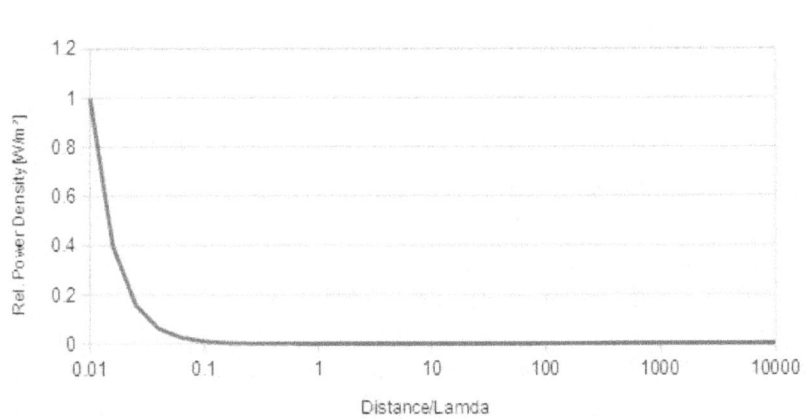

Diagram 1: Power density

Diagram 2 shows the same power density but with double logarithmic scales. Now you can easily find out the density for larger distances.

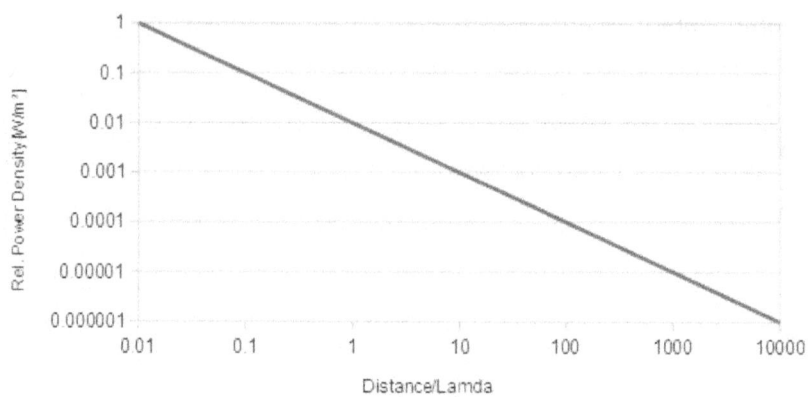

Diagram 2: Double-logarithmic-scaled power density

What is the meaning of the diagram?
1. The largest power density is at the antenna.
2. The power density decreases fast.

To get an impression of which wavelength is at which frequency, look at table 1. These are typical appliances with their frequencies and wavelengths.

Table 1: Wavelength

System	Frequency	Wavelength
Power Line	50 Hz	6000 km
AM long-wave radio	200 kHz	1500 m
AM medium-wave radio	1 MHz	300 m
AM short-wave radio	10 MHz	30 m
FM radio	100 MHz	3 m
UHF channel 38 TV	597.25 MHz	0.5 m
LTE 800 (mobile phone trans.)	800 MHz	375 mm
DECT (cordless) telephone	1890 MHz	159 mm
Microwave oven	2455 MHz	122 mm
WLAN 5.8	5800 MHz	52 mm

Power Delivered

Now you know how the energy is radiated outwards, the next important question is how much radiation will be delivered to an object.

Before we look at the waves, let us see what happens in a circuit:
The circuit consists of the voltage source **U** with its source resistance **Ri** and a load resistance **Rl**.
If we change the load resistance **Rl**, at which value will we get the maximum power delivered to the load resistance?

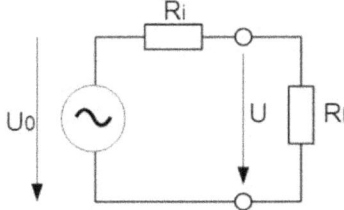

Circuit 1.
For those who want to know, the equations are the following.

$$(1) \quad P = U \cdot I$$
$$(2) \quad U = \frac{U_0 \cdot R_l}{R_i + R_l}$$
$$(3) \quad I = \frac{U_0}{R_i + R_l}$$
$$(4) \quad P = \frac{U_0^2 \cdot R_l}{(R_i + R_l)^2}$$

The power delivered is dependent on the load resistance, which is a non-linear function and not easy to estimate. But for two points we can figure out the results:

1. If we connect a short to the port, this means the load resistance is zero, so the delivered power is also zero.
2. If we leave the port open, this means the load resistance is infinity, so the power delivered is also zero.

Diagram 3 shows the function of the power delivered. The maximum power is delivered when the load resistance is equal to the source resistance.

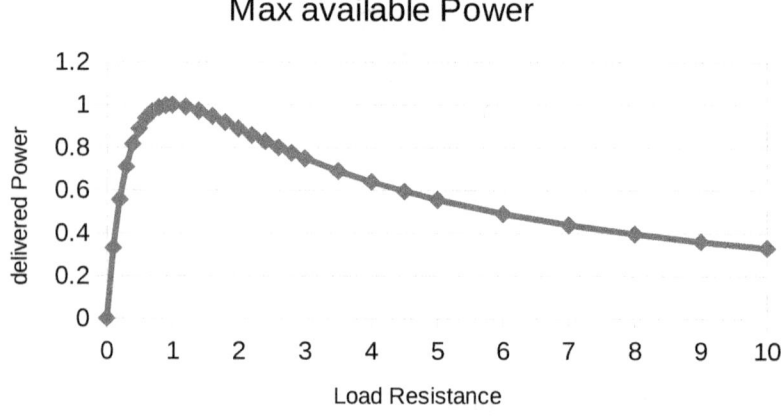

Diagram 3: Delivered power

Let us see what happens with our wave in space. Our wave is far away from the antenna. What is the source impedance for our wave now? It was found it to be around $Z_0 = 377\ \Omega$ or exactly

$$(5)\quad Z_0 = \sqrt{\frac{\mu_0}{\varepsilon_0}} \approx 120 \cdot \pi \cdot \Omega = 377\ \Omega$$

If the wave hits, for example, a water surface perpendicularly, not all the power can be delivered, because water has a different impedance. Its impedance is dependent on its material parameter µ_r and ε_r. With ε_r = 80 and µ_r = 1, its impedance ZF = 42 Ω.

What will happen with the power that is not delivered?

The power not delivered to the water will be reflected. For water, 64% will not be absorbed. Only 36% of the power will travel through the water.

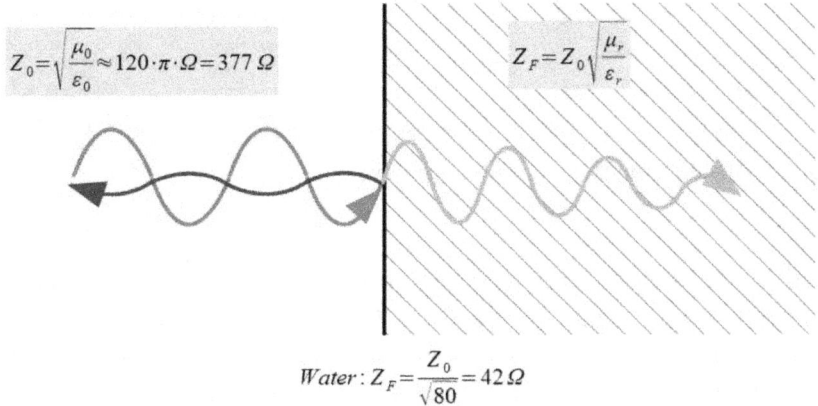

Image 8: *Wave from free space to water.*

Do the electro-magnetic waves affect the water?

Yes. Some part of the energy heats up the water because the water also has a loss. The loss is dependent on the frequency. Diagram 4 shows the permeability and the loss depending on the frequency.

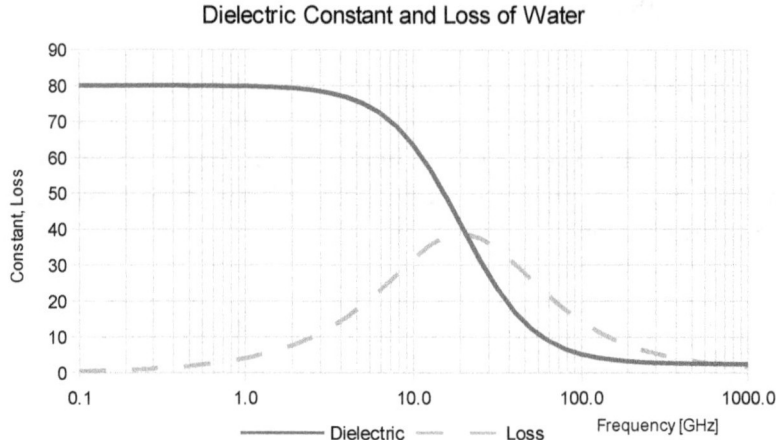

Diagram 4: Permeability and loss of water

The loss in water has is greatest around 20 GHz but a microwave oven works at around 2.45 GHz. The reason is that the loss is dependent on the wave length. At a higher frequency the energy is absorbed in a shorter distance. If we consider a human body, at 20 GHz nearly all energy will be absorbed by the skin. If a microwave oven worked at 20 GHz and we tried to cook a chicken in it, we would get a chicken which has hot skin and cold meat. At 2.45 GHz 99.5% of the energy will be absorbed within 7 cm, and we can be sure our meal will be heated up inside as well.

For industrial applications microwave ovens with 915 MHz are used to heat up larger objects.

Maximum Exposure Level

What implications does all this have when we consider human exposure to microwaves and Wi-Fi, etc?

To protect people from negative long-term effects and accidents, the government has set a specific absorption rate (SAR) limit of 0.08 W/kg for the whole-body average SAR and 2 W/kg for spatial peak SAR. It is difficult to measure the SAR. Normally, the power will be measured with a probe in a dummy but this solution is not practical for most situations.

ARPANSA (Australian Radiation Protection And Nuclear Safety Agency) has recommended a maximum permissible power density [1]. Here you have to measure the E- and H-field and then calculate the power density. This is necessary if you are in the near field of a transmitter. If you are in the far field, only the E- or H-field has to be measured and with the free space impedance of 377 Ω we can calculate the power density. The far field is dependent on the gain of the antenna, current distribution on the antenna and the wavelength. To be sure you should measure at a larger distance than you calculated with equation (6). Hereby G is the antenna gain and λ the wavelength.

$$(6) \quad r \geqslant G \cdot \lambda$$

As I have demonstrated before, the absorption is dependent on the wavelength. At low frequency the wavelength is large compared to a human and the amount of energy absorbed is actually quite low. At high frequency nearly all energy is absorbed by the skin and will definitely not heat the body. To take this circumstance into account, the recommended limit changes with the frequency.

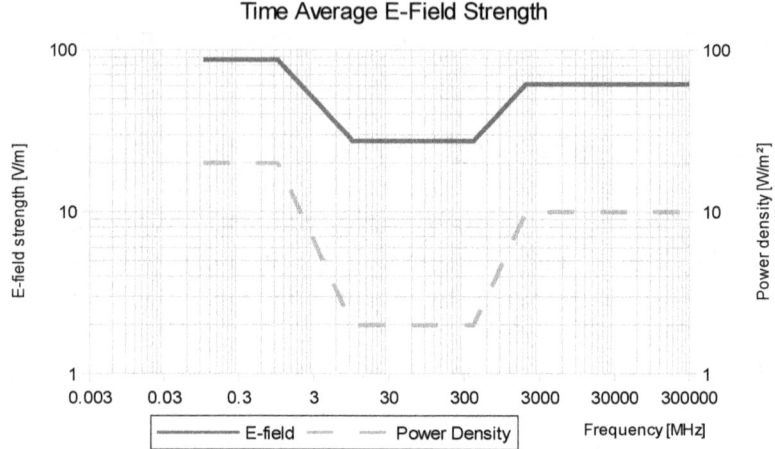

Diagram 5: Maximum exposure level

The ARPANSA limits only take into account thermal effects. In other countries like Russia or Switzerland they recommend up to 100 times lower limits. To be on the safe side, a limit of 0.27V/m at the frequency range 10MHz to 400MHz should not be exceeded at your home, kindergarten, school or office.

Chapter 2
Radiation Systems In Our Lives

Microwave Oven

Picture 1: Microwave oven

The job of a microwave oven is to heat up food. To keep the electro-magnetic field within the oven, every microwave oven has a special door seal. The seal is designed in such a way that frequencies around 2.45 GHz can't travel through the gap between door and oven. Picture 1 shows the set-up for measuring the microwaves from a random oven and the results. As previously explained, we have to measure in the far field and I chose 1.3m. With the relationship of the power density dependent on the distance, we can estimate the field strength near the oven. The power density is around 1W/m² at 10 cm distance and 10% of the recommended limit.

To show the influence of dirt I placed two layers of paper between the door and the oven. The radiated power increased from 6 mW/m² to 30 mW/m² at 1.3m and reached 50% of the limit with 5W/m² at 10cm. A scratch or more dirt could increase the level even more.

Picture 2: Microwave oven with paper

WLAN 2.4 GHz

Picture 3: Wi-Fi modem

In image 9 we see the measurement from a spectrum analyser. The horizontal axis is the frequency, and the vertical axis is the power density in µW/m². This modem transmits the signals on channel 1 and has a maximum power density around 2,418MHz.

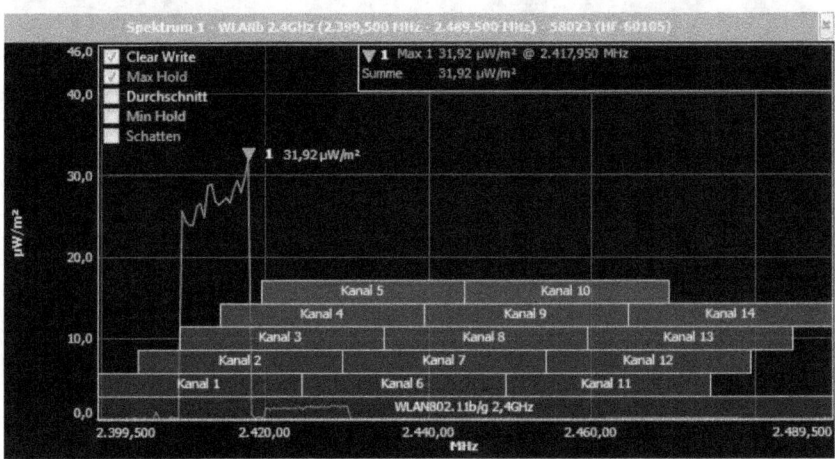

Image 9: WiFi modem at 2.4GHz

The measured level of 31.92 µW/m² is around 0.3% of the limit at a distance of 10 cm.

Cordless Phone

The cordless phone sends a short pulse of around 10 ms and spreads the signal over 10 MHz bandwidth. In the measurement it can be seen that the signal is transmitted from channel 0 to channel 5. The pulse itself is much higher than the spectral density but always under the limit, and is around 0.1% of the limit.

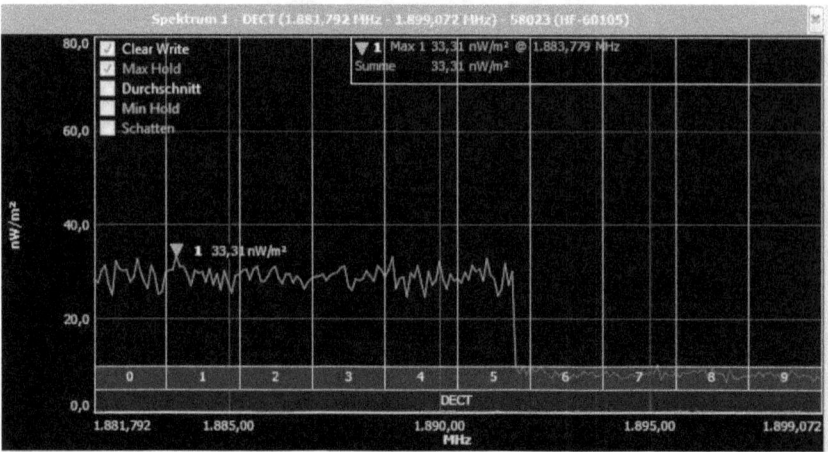

Image 10: Cordless phone

Some people are sensitive to these pulsed signals. If they have a cordless phone at the side of their bed, it could be that they often awake at night. Some "eco" cordless phones stop sending radiation when they are in the base station and could be used if you are sensitive.

Mobile Phone / Cell Phone

Nowadays mobile phones all have internal antennas and it is difficult to design the antenna in such way that the antenna radiates away from the head. But sometimes, depending on the mobile phone standard, GSM800, GSM1800, UMTS, etc, the radiation may be directed **TOWARDS** the head!

In the following diagram of two mobile phones you can see that the antenna radiates in the direction of the head or hand.

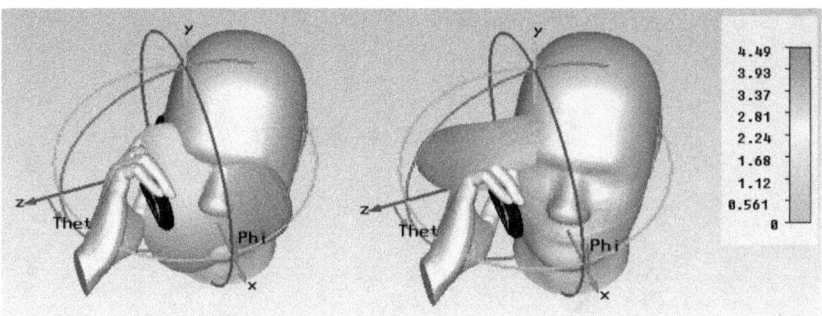

Image 11: Directivity simulation

Now we want to know which SAR (Specific Absorption Rate) we will get if we move the hand away from the head. In image 12 you see the SAR level in the head. The highest level is near the centre of the mobile phone. The SAR level drops down very fast and is 0.7 W/kg at 5 mm distance. It was previously 1.2 W/kg. At 1 cm distance the SAR is only 35% of the maximum value.

The simulation was done with CST and represents a possible solution. This will vary with different mobile phones and hand positions.

Image 12: SAR simulation

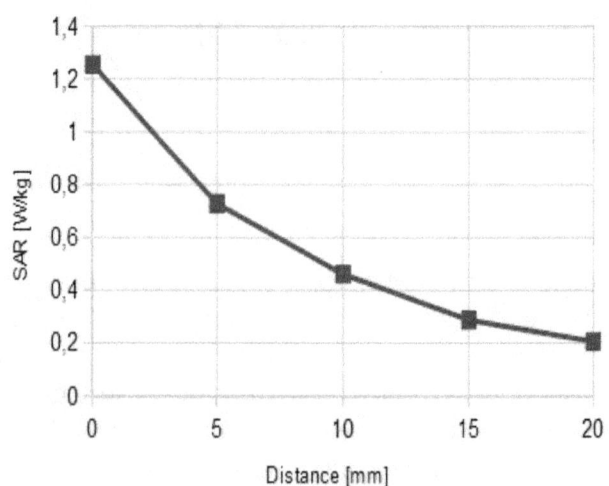

Diagram 6: SAR depending on distance

Smart Meters

There are different smart meters on the market using different frequencies for the wireless connection. These are
915 MHz – 928 MHz
2.3 GHz
2.4 GHz

Each smart meter makes a brief connection to the Back Office two to six times per day. However, any smart meter could be a relay station for the connection to the Back Office. A field survey was conducted by the Victorian government. Because the Smart Meters normally transmit intermittently, for the purpose of the survey, the Smart Meter was set up to transmit continuously.

The measured peak value was between 0.006 and 0.1% of the limit at 30cm distance. The final power density is much less if the Smart Meter works in the normal way.
The distance between the Smart Meters is often much larger in rural areas and some people have reported that measurements have shown that their Smart Meters transmitted with much more power.
It was reported that some older people living in rural areas had problems after their Smart Meter was installed. Measurements apparently showed that the field strength was above the limit. This sounds feasible because the distance between the meters can be large and the technician could have increased the transmitting level without checking if it was above the limit. This is only an assumption and the evidence is purely anecdotal, but worrying none-the-less.

If the Smart Meter is in a metal box, resonance effects could increase the field strength around the box.

Tablets and Notebooks

Apple decided that the iPad can only be connected to the internet via Wi-Fi. Also, most notebooks have the Wi-Fi option and it is handy to use it instead of a cable connection. The problem is, depending on where the antenna is located, a high field strength and a SAR (Specific Absorption Rate) above the limit can occur.

From [2] are some results for SAR (Specific Absorption Rate) exposure when using a notebook where the Wi-Fi antenna is located at different parts of the device. The first two simulations (Image 13) show the SAR exposure if the antenna is placed at the top of the display at two different angles. At an angle of 120° the maximum SAR is 0.06W/kg and within the limit. The first simulation in image 14 is when the antenna is placed at the back of the display. The exposure is far below the limit at 0.005W/kg. The next two simulations show the SAR level of an external Wi-Fi system in use at the USB port. The maximum SAR is 2.742W/kg and over the limit. The exposure is very high at the hands but also at the thighs and abdomen.

A simulation is not available for a tablet or iPad but we can suggest how it would look. The antenna of the iPad is located on the side of the device. This is similar to the external USB Wi-Fi system. If you place it on your lap you will receive a high exposure to your thighs. This is not a position I would recommend. In light of the increased prescribed usage of iPads by school children this radiation is disturbing. Warnings and advice for parents and children on how to use and hold tablets, and strict limits on usage would be commendable.

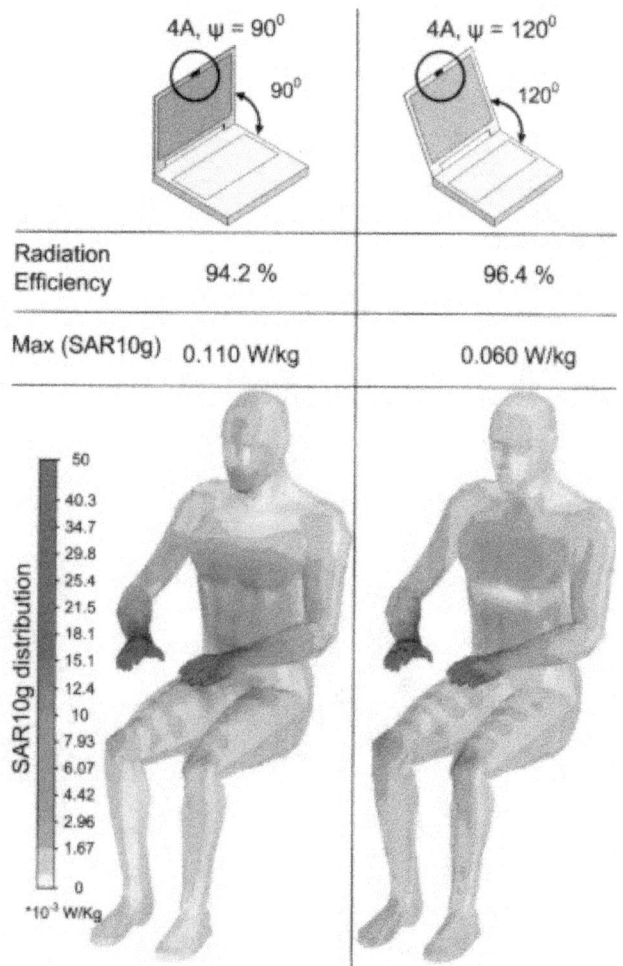

Image 13: Antenna at the front of the display

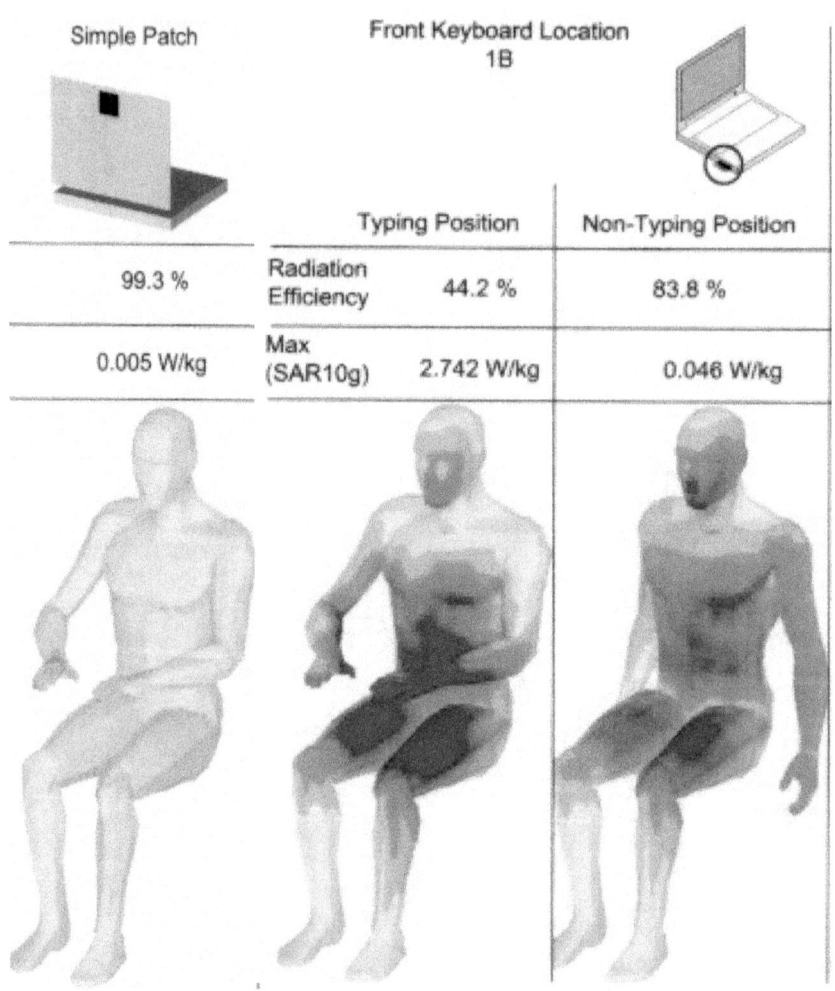

Image 14: Antenna on the display back and front keyboard location

Chapter 3
How To Minimise Exposure

Mobile Phones

As explained before, you get the highest exposure if you call with your mobile phone right next to your head. A small distance leads to a drop in this exposure. Solutions are:
- Use the earphones that come with your mobile phone
- Use the loud-speaker / hands-free option, if your phone has it.
- Use a hand-set as shown in picture 4.

Some companies offer special head-sets which should also prevent guided electro-magnetic waves along the head-set cable. Studies have shown that some guided waves can occur, but with all head-sets the maximum level is 100 times less than holding the phone to your head.

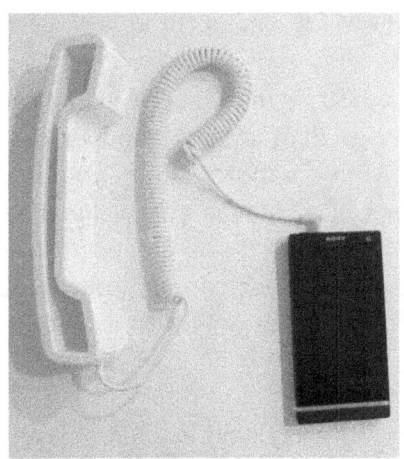
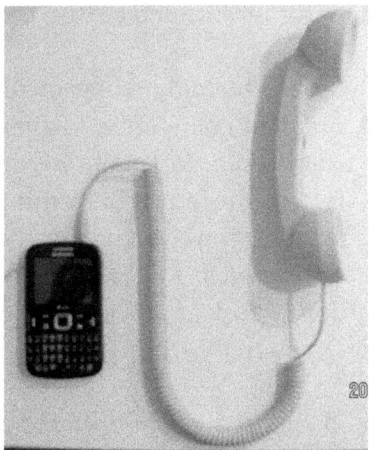

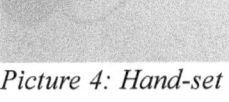
Picture 4: Hand-set

If you have to use your phone without any head-set or hands-free option, keep at least a small distance between your phone and your head. Also, look for a place with good reception, e. g. near a window, and don't walk around during the call.

It is also not advisable to carry your phone in the pocket of your trousers or in a shirt pocket pressed to your body. It is better to place it in either a back-pack or handbag. Even the pocket of your jacket is better.

iPad & Tablet

It is safer not to place an iPad or tablet on your lap. It is better to place it on a table, or on top of a bag or cushion on your lap. Turn off the Wi-Fi if you are not using it.

If you use a USB Wi-Fi system, use a USB extension cable and place the dongle / Wi-Fi stick far away from you. A cable connection to the internet is much better, then the Wi-Fi can be turned off.

If you are building or renovating a house, think about installing a LAN (cable) connection in every room. Plan at least one LAN cable in every room. Normally you have your computer where your desk is. Two or more are better in your office. Maybe you will want to connect a printer or scanner with LAN so that you can use it from everywhere. If you wish to have an internet connection in your garden then add a Wi-Fi modem outside but only turn it on when you need it.

Think about asking your school to minimise iPad use, and turn off the Wi-Fi where practicable.

If you are hyper-sensitive to RF radiation then you can buy a special RF-shielding canopy for above your bed. With an additional shielding carpet under your bed you achieve a very high shielding factor of 10,000 to 1,000,000 depending on the frequency. However, those canopies cost up to $3,000. A similar effect may be achievable with a fine metal mesh, but you have to be aware that you can't use any wireless systems inside.

Microwave Ovens

The gap between the door and the door frame must be clear and clean, without any scratches. Because a microwave oven generates 800W – 1500W of RF energy, a **high radiation level** can occur if the seal is not faultless. If the seal is damaged **DON'T USE** the microwave oven.
To minimise the risk, place the oven away from the usual working area so that people (especially children) can't stand in front of it. Avoid looking into microwave ovens for an extended period of time.

Baby Monitor

A baby monitor should be placed as far as possible from the baby. I don't know if there is a baby monitor with an optional microphone connection but this would help. You could place the microphone near your baby and the baby monitor far away.

Bluetooth Headset and Google Glass

A headset with low energy Bluetooth is an option to minimise your exposure, but the radiation device can be on your head for a long time, and overall the exposure could be greater than phoning with the mobile. In Google Glass there is an integrated Wi-Fi modem. The SAR is under the limit, but the problem is here the same: if you wear it for a long time, the amount will be high.

What doesn't work or is counter-productive:

Cell phone radiation shield stickers
Mobile phone companies spend a lot of money to design the antenna of the phone. They have to achieve a high radiation efficiency with low SAR (Specific Absorption Rate). Their goal is to get the signal into the air and not in the head. If a "shield" sticker or case would have an influence on the radiation, nobody can predict in which way. It could be worse than without it. However, normally a sticker is small and will only have a little influence. Those that are just drawings on paper will have none.

The job of a mobile phone is to radiate the RF energy to transmit the signal to the next base station. It is best not to interfere with this.

Shielding cover
The shielding cover is placed on the back of the mobile phone. It could cause more power to be transmitted through the head instead of away from it.

Shielding pack
Some companies offer a shielding pack into which you put your mobile phone when you are not using it.

Nowadays sophisticated power management is implemented in mobile phones so that the battery will lasts as long as possible. If the reception is good, the phone will transmit at a lower power level to save battery. If the phone is in the shielding pack, it needs higher power to overcome the shielding in order to attempt to keep the connection to the base station and the battery will became flat at a faster rate.

Plug-in anti-radiation devices and stickers
Some merchandisers are promoting the advantages of a plug-in device to neutralise radiation: this cannot possibly work! If you have read the first chapter, you will also understand that stickers attached to the wall or elsewhere may make you feel better or more relaxed and the salesperson happy and richer, but will have no effect on electro-magnetic radiation.

In conclusion: I am against fear-mongering, and unscrupulous capital gain from a worried public. However, I do recommend checking or replacing old microwaves, taking metal doors off smart meter boxes, and getting the iPad out of your lap and mobile phone away from your head.

Bibliography

[1] Maximum Exposure Levels to Radiofrequency Fields — 3kHz to 300GHz; Radiation Protection Series No. 3; Australian Radiation Protection and Nuclear Safety Agency, April 2002

[2] Wrapped Microstrip Antennas for Laptop Computers; J. Guterman, A. Moreira, C. Peixeiro, Y. Rahmat-Samii; IEEE Antenna & Propagation Magazine, Vol. 51, No. 4, August 2009

[3] Antennen und Strahlungsfelder; Klaus W. Kark; Vieweg + Teubner, 2010

www.ingramcontent.com/pod-product-compliance
Lightning Source LLC
Chambersburg PA
CBHW050252230526
45470CB00005B/2221